MW01107132

# An Illustrated Data Guide To
# Modern Artillery

Compiled by
Christopher Chant

TIGER BOOKS INTERNATIONAL
LONDON

This edition published in 1997 by
Tiger Books International PLC
Twickenham

Published in Canada in 1997 by
Vanwell Publishing Limited
St. Catharines, Ontario



Printed in Hong Kong

ISBN 1-85501-861-6

# CONTENTS

# **N**ORINCO **T**ype 83 152mm **SP** gun/howitzer

**Country of origin:** China

**Type:** Tracked self-propelled gun/howitzer

**Crew:** Five

**Combat weight:** 66,138lb (30,000kg)

**Dimensions:** Length, gun forward 24ft 0.6in (7.33m) and hull 22ft 7in (6.882m); width 10ft 7.5in (3.236m); height 11ft 6in (3.502m)

**Armament system:** One 152mm (6in) modified Type 66 gun/howitzer with 30 rounds of ready-use ammunition, one 12.7mm (0.5in) Type 54 AA heavy machine-gun with 650 rounds, one 7.62mm (0.3in) co-axial machine-gun with 650 rounds, and one Type 40 rocket-launcher with four rounds in a hydraulically powered turret; direct- and indirect-fire sights are fitted

**Armour:** Welded steel

**Powerplant:** One 523hp (390kW) Type 12150L diesel engine with 195 Imp gal (885 litres) of internal fuel

**Performance:** Speed, road 34.2mph (55km/h) and cross-country 21.75mph (35km/h); range, road 280 miles (450km); fording 4ft 3.25in (1.30m) without preparation; gradient 32%; side slope 30%; vertical obstacle 2ft 3.6in (0.70m); trench 8ft 10.25in (2.70m); ground clearance 1ft 5.7in (0.45m)

**Variant**

**Type 83:** First revealed in 1984 although it entered service with the Chinese army some years before this, the Type 83 is a useful piece of self-propelled ordnance by the standards prevailing in East and South-East Asia. The hull is a purpose-designed unit also used by a 122mm (4.8in) self-propelled multiple rocket-launcher, and is notable for having the four central units of its six road wheels grouped in pairs; the drive sprocket is at the rear, the idler is at the front, and there are three track-return rollers.

The ordnance appears to be a derivative of the Type 66 towed gun/howitzer fitted with a fume extractor, and can be elevated in an arc of 70° between -5° and +65°, while the turret can be power-traversed through 360°. Assuming that the ordnance of the self-propelled equipment has the same performance as the Type 66, the Type 83 can fire HE and smoke projectiles, each weighing 96.03lb (43.56kg) and possessing filling of 12.92lb (5.86kg) of TNT and 14.61lb (6.627kg) of white phosphorus respectively, at a muzzle velocity of 2,149ft (655m) per second for a maximum range of 18,845yds (17,230m). The Type 66 can also fire a 96.03lb (43.56kg) rocket-assisted projectile to a range of 23,930yds (21,880m) without any sacrifice of accuracy, it is claimed, and it is probable that the ordnance of the self-propelled equipment can also fire this MP-152 projectile. The maximum rate of fire is four rounds per minute.

*The Chinese Type 83 is a moderately advanced equipment using a chassis specifically designed for this application rather than an adaptation of a main battle tank chassis, and hydraulic power is used for traverse of the turret and elevation of the ordnance. There is no power-assisted loading system, and only a moderately advanced fire-control system is incorporated.*

# Tatra DANA 152mm SP howitzer

**Country of origin:** Czechoslovakia (now Czech Republic)

**Type:** Wheeled self-propelled howitzer

**Crew:** Probably three

**Combat weight:** 50,705lb (23,000kg)

**Dimensions:** Length, gun forward 34ft 1.5in (10.40m) and hull 29ft 1.25in (8.87m); width, turret 9ft 9in (2.97m) and hull 8ft 11.25in (2.722m); height to top of machine-gun 11ft 6.75in (3.525m)

**Armament system:** One 152mm (6in) 2S3 (Model 1973) howitzer with an unrevealed number of rounds of ready-use ammunition, and one 12.7mm (0.5in) DShKM AA heavy machine-gun in a hydraulically powered turret; direct- and indirect-fire sights are fitted

**Armour:** Welded steel

**Powerplant:** One 349hp (260kW) Tatra T-928 (T3-930) multi-fuel engine with an unrevealed quantity of internal fuel, and driving an 8 x 8 layout

**Performance:** Speed, road 49.7mph (80km/h); range, road 621 miles (1,000km); fording 4ft 7.5in (1.40m) without preparation; gradient 60%; side slope not revealed; vertical obstacle not revealed; trench 6ft 6.75in (2.00m); ground clearance not revealed

**Variant**

**DANA:** This very uselul SP howitzer was introduced into service in 1981, and is based on the chassis and automotive system of the Tatra 815 8 x 8 high-mobility truck, which offers the same type of cross-country mobility as a tracked chassis but at considerably less cost. The DANA has a central tyre pressure-regulation system so that the best pressure can be selected for each type of terrain encountered.

The howlizer crew and the ammunition are carried in a separate vehicle, and for a higher rate of fire the ordnance has an automatic loading system powered by a hydraulic motor which also traverses the turret through 360° and provides power for elevation in an arc of 63° bdween -3° and +60° In firing position the vehicle is stabilised by three hydraulically actuated jacks, the location of these (one at each side and the third at the rear) suggesting that the DANA may be designed to fire only over the frontal arc despite the 360° traverse available to the turret, which is located centrally and accommodates the ordnance in a central slot that removes the need for a bore evacuator, as the members of the cww are accommodated in the two compartments to the sides of the ordnance.

The ordnance is based on the 8oviet 283 weapon and fires its primary projectile, namely the 95.92lb (43.51kg) OF-540 HE-FRAG type, to a range of 1 9,040yds (17,41 0m). Other ammunition types are an HE rocket-assisted projectile fired to a range of some 26,245yds (24,000m), the 107.58lb (48.8kg) AP HE projectile, the S-540 illuminating projectile and the D-540 smoke projectile.

*The use of a wheeled chassis keeps down the cost of the DANA and also provides useful range and high road speed.The turret has full traverse via a power system also used for ordnance elevation, and the central location of the turret helps balance the vehicle in firing position.*

# GIAT GCT 155mm SP gun

**Country of origin:** France

**Type:** Tracked self-propelled gun

**Crew:** Four

**Combat weight:** 92,593lb (42,000kg)

**Dimensions:** Length, gun forward 33ft 7.5in (10.25m) and hull 21ft 11.75in (6.70m); width 10ft 4in (3.15m); height to turret top 10ft 8in (3.25m)

**Armament system:** One 155mm (6.1in) GIAT gun with 42 rounds of ready-use ammunition, one 7.62mm (0.3in) or 0.5in (12.7mm) AA machine-gun with 2,050 or 800 rounds respectively, and two smoke-dischargers on the front of the hydraulically powered turret; direct- and indirect-fire sights are fitted

**Armour:** Welded steel

**Powerplant:** One 718hp (535kW) Hispano-Suiza HS 110 multi-fuel engine with 213 Imp gal (970 litres) of internal fuel

**Performance:** Speed, road 37.3mph (60km/h); range, road 280 miles (450km); fording 6ft 10.67in (2.10m) without preparation; gradient 60%; side slope 30%; vertical obstacle 3ft 0.6in (0.93m); trench 6ft 3in (1.90m); ground clearance 1ft 4.5in (0.42m)

**Variant**

**GCT:** Designed to replace obsolescent 105 and 155mm (4.13 and 6.1in) self-propelled weapons based on the chassis of the AMX-13 light tank, the GCT (Grande Cadence de Tir, or great rate of fire) entered service in 1978, initially with the army of Saudi Arabia and then with the French army (under the designation 155 AU F1) and Iraq. The equipment is based on the chassis of the AMX-30 main battle tank (MBT) and therefore shares the same level of battlefield mobility while reducing logistical and maintenance requirements.

*The GCT has good gun performance as a result of its automatic loading system, and possesses full tactical compatibility with the AMX-30 main battle tank as it uses the same chassis.*

The type replaces the MBT's low-silhouette turret with a voluminous unit accommodating the ordnance, ammunition, and an automatic loading system that permits a high rate of fire using the onboard ammunition supply (a burst capability of six rounds in 45 seconds is possible). Reloading of the ammunition supply can be undertaken while the gun is in action, four men accomplishing a complete reload in 15 minutes. Optional equipment includes an NBC (nuclear, biological, chemical) protection system and specialised artillery items.

The L/40 ordnance can be elevated through an arc of 70° (-4° to +66°) in the turret, which can be traversed through 360°, and fires a range of ammunition types based on a combustible cartridge case that is wholly consumed in the firing process to leave no case for extraction before the fresh round is loaded. Projectile types are HE-FRAG, HE rocket-assisted, anti-tank mine, smoke and illuminating, and the typical range is 23,185yds (21,200m) with the 96.45lb (43.75kg) OE 155 56/59 HE-FRAG projectile that contains a 19.62lb (8.9kg) HT 50/50 explosive filling, although the 93.7lb (42.5kg) OE PAD 155 H3 HE RAP with a 19.18lb (8.7kg) HT 50/50 explosive filling attains a range of 33,355yds (30,500m). The 101.41lb (46kg) OMI 155 H2 cargo round carries six 3.97lb (1.8kg) anti-tank mines to a maximum range of 19,685yds (18,000m), and the 95.9lb (43.5kg) OE DTC 155 H2 HE base-bleed round carries its 19.18lb (8.7kg) HT 50/50 explosive filling to a maximum range of 31,170yds (28,500m). The ordnance can also fire the US Copperhead cannon-launched projectile, which is a highly capable anti-tank weapon with semi-active laser guidance.

# Creusot-Loire Modèle F3 155mm SP gun/howitzer

**Country of origin:** France

**Type:** Tracked self-propelled gun/howitzer

**Crew:** Two

**Combat weight:** 38,360lb (17,400kg)

**Dimensions:** Length 20ft 5in (6.22m); width 8ft 11in (2.72m); height 6ft 10in (2.085m)

**Armament:** One 155mm (6.1in) ATS gun/howitzer with no ready-use ammunition on a limited-traverse mounting; direct- and indirect-fire sights are fitted

**Armour:** Welded steel varying in thickness from 0.39 to 0.79in (10 to 20mm)

**Powerplant:** One 248hp (185kW) SOFAM 8Gbx petrol engine or one 280hp (209kW) Detroit Diesel 6V-53T diesel engine with 99 Imp gal (450 litres) of internal fuel

**Performance:** Speed, road 37.3mph (60km/h) with petrol engine or 40mph (64km/h) with diesel engine; range, road 186 miles (300km) with petrol engine or 249 miles (400km) with diesel engine; fording 3ft 3.5in (1.00m)

*The Mk F3 is now completely obsolete as a result mainly of the fact that its ordnance, itself an elderly type, is completely exposed and its crew are therefore highly vulnerable to a multitude of air- and ground-launched anti-personnel weapons as well as more insidious attack from nuclear, biological and chemical agents.*

without preparation; gradient 40%; side slope not revealed; vertical obstacle 1ft 11.6in (0.60m); trench 4ft 11in (1.50m); ground clearance 1ft 7in (0.48m)

**Variant**

**Modèle F3:** This self-propelled gun is based on the shortened chassis of the AMX-13 light tank, and after its introduction in the late 1950s was offered in petrol- and diesel-engined models, with kits available to permit the operators of petrol-engined versions to convert their equipments to diesel power.

The ordnance is located on an open mounting at the rear of the vehicle, and can here be traversed through a total of 50° (20° left and 30° right of the centreline) at elevation angles between 0° and +50°, and a total of 46° (16° left and 30° right of the centreline) at elevation angles between +50° and +67°; the total elevation arc is thus 67°. The vehicle is stabilised while firing by twin spades at the rear.

The eight-man gun crew and 25 rounds of ready-use ammunition are carried in an accompanying VCA tracked vehicle. The L/33 ordnance fires HE-FRAG, HE rocket-assisted, smoke and illuminating projectiles, the 96.45lb (43.75kg) HE-FRAG type leaving the muzzle at a velocity of 2,379ft (725m) per second to reach a maximum range of 21,925yds (20,050m).

Operators of the Modèle 51 have included Argentina, Chile, Ecuador, France, Kuwait, Morocco, Qatar, Sudan, United Arab Emirates, and Venezuela, and the equipment still serves with many of these countries despite its basic technical obsolescence and lack of any protection for the gun and its crew.

# Creusot-Loire Modèle 61 105mm SP howitzer

**Country of origin:** France

**Type:** Tracked self-propelled howitzer

**Crew:** Five

**Combat weight:** 36,375lb (16,500kg)

**Dimensions:** Length 18ft 8.5in (5.70m); width 8ft 8.25in (2.65m); height to top of cupola 8ft 10.25in (2.70m)

**Armament system:** One 105mm (4.13in) Modèle 50 howitzer with 56 rounds of ready-use ammunition on a manually powered limited-traverse mounting, and one or two 7.5 or 7.62mm (0.295 or 0.3in) machine-guns (one for AA defence and one optional for local defence) with 2,000 rounds; direct- and indirect-fire sights are fitted

**Armour:** Welded steel varying in thickness from 0.39 to 0.79in (10 to 20mm)

**Powerplant:** One 248hp (185kW) SOFAM 8Gbx petrol engine with 91 Imp gal (415 litres) of internal fuel

**Performance:** Speed, road 37.3mph (60km/h); range, road 217 miles (349km); fording 2ft 7.5in (0.80m) without preparation; gradient 60%; side slope 30%; vertical obstacle 2ft 1.6in (0.65m); trench 5ft 3in (1.60m); ground clearance 10.8in (0.275m)

**Variant**

**Modèle 61:** Designed in the late 1940s, the Modèle 61 is now obsolescent in terms of its chassis, which was derived from that of the AMX-13 light tank, and of its ordnance. The equipment was accepted for French service in 1958, and was also sold to Indonesia and Morocco. By modern battlefield standards, the type is limited by its lack of NBC protection and amphibious capability.

The ordnance can be traversed through a total of only 40° (20° left and right of the centreline) through an elevation arc of 70.5° (-4.5° to +66°) on its mounting in the rear-mounted armoured fighting compartment. The L/23 or L/30 ordnance fires HE-FRAG or HEAT projectiles, the former weighing 35.3lb (16kg) and attaining a maximum range of 16,405yds (15,000m) after being fired at a muzzle velocity of 2,198ft (670m) per second, and the latter being capable of penetrating a maximum of 13.78in (350mm) of armour at any range after being fired at a muzzle velocity of 2,297ft (700m) per second.

*Based on the chassis of the AMX-13 light tank, the Mk 61 lacks NBC protection and amphibious capability. The equipment also has a fixed fighting compartment rather than a traversing turret, and while this offers the crew protection against small-arms fire and light fragmentation weapons, it also requires the whole vehicle to be slewed for gross changes in azimuth.*

# Soltam M-72 155mm SP gun/howitzer

**Country of origin:** Israel

**Type:** Tracked self-propelled gun/howitzer

**Crew:** Five

**Combat weight:** Not revealed

**Dimensions:** Length, gun forward not revealed, hull 25ft 8in (7.823m); width 11ft 1.5in (3.39m); height not revealed

**Armament system:** One 155mm (6.1in) Soltam gun/howitzer with an unrevealed number of rounds of ready-use ammunition in a hydraulically operated turret, and one 0.5in (12.7mm) Browning M2HB AA heavy machine-gun with an unrevealed number of rounds, or one 7.62mm (0.3in) AA machine-gun with an unrevealed number of rounds; direct- and indirect-fire sights are fitted

**Armour:** Welded steel varying in thickness between 0.67 and 3in (17 and 76.2mm)

**Powerplant:** One 750hp (559kW) Teledyne Continental Motors AVDS-1790-2A diesel engine with an unrevealed quantity of internal fuel

**Performance:** Speed, road 26.7mph (43km/h); range, road about 373 miles (600km); fording 4ft 9in (1.45m) without preparation; gradient 60%; side slope 30%; vertical obstacle 3ft 0in (0.91m); trench 11ft 0in (3.35m); ground clearance 1ft 8in (0.51m)

**Variant**

**M-72:** This may be regarded as the modern equivalent of the L-33 equipment, using a similar L/33 ordnance although the type is also offered with a more modern L/39 ordnance. The hull is basically that of the Centurion, an obsolete British MBT used in large numbers by the Israeli army, in a form updated and improved by the Israelis with a diesel-engined powerplant and other features, including the relocation of the driver to the right of the hull front allowing the fitment of a large left-hand door through which ammunition can be loaded even as the equipment is in action.

The comparatively small turret weighs 30,864lb (14,000kg) complete with ammunition, and in action is generally traversed to fire over the rear of the vehicle. This turret can be traversed through 360°, and the ordnance elevated through an arc of 68° (-3° to +65°). A moderately high rate of fire is made possible by the use of a pneumatically operated rammer that allows loading at all angles of elevation, and the L/33 ordnance uses standard NATO types of ammunition, including a 96.34lb (43.7kg) HE-FRAG projectile fired at a muzzle velocity of 2,379ft (725m) per second to a maximum range of 22,420yds (20,500m): the equivalent figures for the L/39 ordnance with the same projectile are a muzzle velocity of 2,690ft (820m) per second and a maximum range of 25,700yds (23,500m).

The complete equipment is offered with a number of optional features including full NBC protection, and the turret can also be installed on a number of other MBT chassis.

*Although a somewhat ungainly equipment in appearance as a result of its massive turret on the chassis of the elderly Centurion main battle tank, the M-72 is a durable and highly reliable equipment with good cross-country mobility as well as a powerful ordnance.*

# Soltam L-33 155mm SP gun/howitzer

**Country of origin:** Israel

**Type:** Tracked self-propelled gun/howitzer

**Crew:** Eight

**Combat weight:** 91,490lb (41,500kg)

**Dimensions:** Length, gun forward 27ft 9.5in (8.47m) and hull 21ft 2.75in (6.47m); width 11ft 5.75in (3.50m); height 11ft 3.75in (3.45m)

**Armament system:** One 155mm (6.1in) Soltam M-68 gun/howitzer with 60 rounds of ready-use ammunition on a manually powered limited-traverse mounting, and one 7.62mm (0.3in) AA machine-gun with an unrevealed number of rounds; direct- and indirect-fire sights are fitted

**Armour:** Welded steel varying in thickness between 0.63 and 2.5in (16 and 63.5mm)

**Powerplant:** One 460hp (343kW) Cummins VT8-460-Bi diesel engine with 180 Imp gal (820 litres) of internal fuel

**Performance:** Speed, road 23.6mph (38km/h); range, road 162 miles (260km); fording 3ft 0in (0.91m) without preparation; gradient 60%; side slope 30%; vertical obstacle 3ft 0in (0.91m); trench 7ft 6in (2.30m); ground clearance 1ft 5in (0.43m)

**Variant**

**L-33:** This simple yet effective Israeli development locates an M-68 gun/howitzer in a massive fixed fighting compartment on the chassis of an obsolete M4A3E8 Sherman medium tank of the type with horizontal-volute rather than vertical-volute spring suspension. The ordnance is installed in a tall slab-sided fighting compartment built upwards from the basic hull of the vehicle, and can be traversed through a total of 60° on its mounting and through an arc of 55 (-3° to +52°).

The type has proved operationally successful since it entered Israeli service in 1973; the L/33 ordnance can use all types of NATO ammunition in its calibre, including a 96.3lb (43.7kg) HE-FRAG projectile fired at a muzzle velocity of 2,379ft (725m) per second to a maximum range of 21,870yds (20,000m). The equipment was produced in moderately large numbers for the Israeli army, with whose reserves it still serves in modest numbers, but no export sales were achieved.

*Based on an old chassis, that of the M4 Sherman medium tank of World War II, the L-33 is still a useful equipment, with powerful ordnance located in the massive superstructure built up from the chassis as a blocky unit offering good protection against light and medium battlefield weapons.*

# EEFA Bourges M-50 155mm SP howitzer

**Country of origin:** France (to an Israeli requirement)

**Type:** Tracked self-propelled howitzer

**Crew:** Eight

**Combat weight:** 68,342lb (31,000kg)

**Dimensions:** Length 20ft 0.125in (6.10m); width 9ft 9.25in (2.98m); height 9ft 2.25in (2.80m)

**Armament system:** One 155mm (6.1in) Modèle 50 howitzer with an unrevealed number of rounds of ready-use ammunition on a manually operated limited-traverse mounting; direct- and indirect-fire sights are fitted

**Armour:** Welded steel

**Powerplant:** One 450hp (336kW) Ford GAA petrol engine with 140 Imp gal (636 litres) of internal fuel

**Performance:** Speed, road about 28mph (45km/h); range, road 100 miles (161km); fording 3ft 0in (0.91m) without preparation; gradient 60%; side slope not revealed; vertical obstacle 2ft 0in (0.61m); trench 7ft 5in (2.26m); ground clearance 1ft 5in (0.43m)

**Variant**

**M-50:** This self-propelled howitzer was developed in France during the 1950s to meet an Israeli requirement, and combines a French Modèle 50 howitzer with an open-topped fighting compartment built onto the chassis and lower hull of an M4 Sherman medium tank of the type with vertical-volute rather than horizontal-volute spring suspension, reworked in this instance so that the engine is located in the front rather than at the rear of the hull, thereby freeing space for the installation of the ordnance.

The maximum elevation of the rear-mounted ordnance is +69°, and this howitzer can fire a number of ammunition types including the 94.8lb (43kg) HE projectile, which reaches a maximum range of 19,250yds (17,600m) after being fired at a muzzle velocity of 2,133ft (650m) per second.

The M-50 was produced only for Israel, and is now completely obsolete as a result of its open-topped fighting compartment and other outdated features.

*The Modèle 50 howitzer was developed as a 6.1in (155mm) towed weapon in the period after World War II, and is an obsolete weapon in its basic role. The same ordnance is used in the obsolescent EEFA Bourges M-50 self-propelled howitzer designed in France to meet an Israeli requirement.*

# OTO Melara Palmaria 155mm SP howitzer

**Country of origin:** Italy

**Type:** Tracked self-propelled howitzer

**Crew:** Five

**Combat weight:** 101,411lb (46,000kg)

**Dimensions:** Length, gun forward 37ft 7.75in (11.474m) and hull 24ft 3.33in (7.40m); width 11ft 1.33in (3.386m); height without machine-gun 9ft 5in (2.874m)

**Armament system:** One 155mm (6.1in) OTO Melara L/41 howitzer with 30 rounds of ready-use ammunition, and one 0.5in (12.7mm) or 7.62mm (0.3in) AA machine-gun with an unrevealed number of rounds in a hydraulically powered turret; direct- and indirect-fire sights are fitted

**Armour:** Welded steel and aluminium

**Powerplant:** One 718hp (535kW) MTU MB 837 Ea diesel engine with 198 Imp gal (900 litres) of internal fuel

**Performance:** Speed, road 37.3mph (60km/h); range, road 249 miles (400km); fording 3ft 11.25in (1.20m) without preparation and 13ft 1.5in (4.00m) with preparation; gradient 60%; side slope 30%; vertical obstacle 3ft 3.5in (1.00m); trench 9ft 10in (3.00m); ground clearance 1ft 3.75in (0.40m)

**Variant**

**Palmaria:** This is the chassis of the OF-40 MBT with a lower-powered engine and modifications to accommodate a larger turret and an OTO Melara howitzer with an automatic loader permitting a rate of fire of three rounds in 20 seconds, or one round per minute for 1 hour, or one round every 3 minutes indefinitely. The turret can be traversed through 360°, and the ordnance can be elevated in an arc of 74° (-4° to +70°). The ordnance can fire a large assortment of NATO standard ammunition types, including the 95.9lb (43.5kg)

*The Palmaria is a modern equipment offering many advanced features including a capable ordnance, yet it is still an affordable item.*

HE-FRAG projectile fired to a maximum range of 27,010yds (24,700m) or the 95.9lb (43.5kg) HE rocket-assisted projectile to a maximum range of 32,810yds (30,000m); there are also smoke and illuminating projectile types.

The ordnance can also fire a range of ammunition developed specifically for this equipment by Simmel in Italy, although this ammunition is capable of being fired by other 155mm (6.1in) NATO artillery equipments. The HE type is the 95.9lb (43.5kg) P3, which is available in three forms: as the basic P3 with a 25.8lb (11.7kg) explosive charge and a maximum range of 26,245yds (24,000m); as the long-trajectory P3 with the same explosive filling but with a maximum range of 30,075yds (27,500m) through the adoption of a number of drag reduction features; and as the P3 rocket-assisted type with the explosive filling reduced to 17.64lb (8kg) but with the range increased to 32,810yds (30,000m); there are also P4 illuminating and P5 smoke projectiles.

The Palmaria is in service with the Italian army, and has also been ordered for Argentina (Palmaria turret on the chassis of the TAM medium tank), Libya and Nigeria.

# Mitsubishi Type 75 155mm SP howitzer

**Country of origin:** Japan

**Type:** Tracked self-propelled howitzer

**Crew:** Six

**Combat weight:** 55,776lb (25,300kg)

**Dimensions:** Length, gun forward 25ft 6.67in (7.79m) and hull 21ft 9.5in (6.64m); width 10ft 1.67in (3.09m); height to turret roof 8ft 4.25in (2.545m)

**Armament system:** One 155mm (6.1in) NSJ howitzer with 28 rounds of ready-use ammunition and one 0.5in (12.7mm) Browning M2HB AA heavy machine-gun with 1,000 rounds in a hydraulically powered turret; direct- and indirect-fire sights are fitted

**Armour:** Welded aluminium

**Powerplant:** One 449hp (335kW) Mitsubishi 6ZF diesel engine with 143 Imp gal (650 litres) of internal fuel

**Performance:** Speed, road 29.2mph (47km/h); range, road 186 miles (300km); fording 4ft 3in (1.30m) without preparation; gradient 60%; side slope 30%; vertical obstacle 2ft 3.5in (0.70m); trench 8ft 2.5in (2.50m); ground clearance 1ft 3.75in (0.40m)

**Variant**

**Type 75:** This Japanese self-propelled howitzer entered service in 1978, and is essentially a simple equipment that resembles the US M109 series in layout. The L/30 ordnance is located in a rear-mounted turret that can traverse through 360° and permits ordnance elevation in an arc of 75° (-5° to +70°). The ordnance can fire a number of projectile types, including HE to a maximum range of 20,770yds (18,990m), and HE rocket-assisted to a maximum range of 26,245yds (24,000m).

The type is used only by Japan, and has an NBC protection system.

*The Type 75 is in no way an exceptional equipment, but is adequate for the purposes of the Japanese Ground Self-Defence Force with its good ordnance, full protection and an NBC system.*

# Komatsu/Japan Steel Works Type 74 105mm SP howitzer

**Country of origin:** Japan

**Type:** Tracked self-propelled howitzer

**Crew:** Five

**Combat weight:** 36,376lb (16,500kg)

**Dimensions:** Length 19ft 4.25in (5.90m); width 9ft 6in (2.90m); height 7ft 10in (2.39m)

**Armament system:** One 105mm (4.13in) NSJ howitzer with an unrevealed number of ready-use rounds and one 0.5in (12.7mm) Browning M2HB AA heavy machine-gun with an unrevealed number of rounds in a hydraulically powered turret; direct- and indirect-fire sights are fitted

**Armour:** Welded steel

**Powerplant:** One 302hp (225kW) Mitsubishi 4ZF diesel engine with an unrevealed quantity of internal fuel

**Performance:** Speed, road 31mph (50km/h); range, road 186 miles (300km); fording 3ft 11.25in (1.20m) without preparation and amphibious with preparation; gradient 60%; side slope not revealed; vertical obstacle 2ft 0in (0.61m); trench 6ft 7in (2.00m); ground clearance 1ft 3.75in (0.40m)

**Variant**

**Type 74:** Based on the chassis and automotive system of the Type 73 armoured personnel carrier, the Type 74 self-propelled howitzer entered service in 1975. The rear-mounted turret traverses through 360°, but no details of ammunition types or ordnance have been revealed. It is thought that the type, which is used only by Japan, has an NBC protection system and other advanced features.

# Denel (Armscor) G6 Renoster 155mm SP howitzer

**Country of origin:** South Africa

**Type:** Wheeled self-propelled howitzer

**Crew:** Six

**Combat weight:** 103,263lb (46,840kg)

**Dimensions:** Length, gun forward 33ft 11in (10.335m) and hull 30ft 2in (9.20m); width 11ft 1.75in (3.40m); height to turret top 10ft 6in (3.20m)

**Armament system:** One 155mm (6.1in) Armscor G5 howitzer with 47 rounds of ready-use ammunition, one 0.5in (12.7mm) Browning M2HB AA heavy machine-gun with 900 rounds, and four 81mm (3.2in) grenade-launchers on each side of the hydraulically powered turret; direct- and indirect-fire sights are fitted

**Armour:** Welded steel

**Powerplant:** One 525hp (391kW) diesel engine with 154 Imp gal (700 litres) of fuel, driving a 6 x 6 layout

**Performance:** Speed, road 56mph (90km/h); range, road 373 miles (600km); fording 3ft 3.5in (1.00m) without preparation; gradient 50%; side slope 30%; vertical obstacle 1ft 5.75in (0.45m); trench 3ft 3.5in (1.00m); ground clearance 1ft 5.75in (0.45m)

**Variant**
**G6 Renoster:** This self-propelled howitzer, which entered service in the mid-1980s and whose name means rhinoceros in Afrikaans, was inspired by a Canadian idea but realised entirely in South Africa as a long-range weapon able to support other armoured vehicles in all types of terrain. The equipment is based on a substantial wheeled chassis with 6 x 6 drive, and carries a rear-mounted turret capable of 360° traverse and carrying the advanced and powerful L/45 ordnance that can be elevated through an arc of 80° (-5° to +75°).

*The ordnance carried in the G6 Renoster wheeled self-propelled equipment is the same as that used by the 6.1in (155mm) G5 towed gun/howitzer. This is a truly exceptional ordnance able to fire a wide assortment of projectiles with very considerable accuracy to very long range.*

The ordnance fires the M57 family of extended-range full-bore projectiles developed for the G5 towed howitzer, and this family now includes a combustible cartridge case that facilitates a rapid case extraction and reloading cycle for the maximum rate of re together with the minimum clutter of the ghting compartment. The ammunition family includes HE-FRAG, HE base-bleed, HE rocket-assisted, smoke and illuminating projectiles. The HE-FRAG projectile weighs 100.3lb (45.5kg) with a 19.18lb (8.7kg) HE lling and reaches a maximum range of 32,810yds (30,000m), while the 103.6lb (47kg) HE base-bleed projectile with the same explosive lling attains a highly capable 43,745yds (40,000m). A cargo round is known to have been developed for the delivery of anti-tank mines to a considerable range, but no details of this projectile have been released.

# Fabrica de Artilleria de Sevilla SB 155/39 ATP 155mm SP howitzer

**Country of origin:** Spain

**Type:** Tracked self-propelled howitzer

**Crew:** Five

**Combat weight:** 83,774lb (38,000kg)

**Dimensions:** Length, gun forward 32ft 2in (9.80m), hull not revealed; width 10ft 6in (3.20m); height to turret top not revealed

**Armament system:** One 155mm (6.1in) Santa Barbara 155/39 howitzer with 28 rounds of ready-use ammunition, and one 0.5in (12.7mm) Browning M2HB

AA heavy machine-gun with an unrevealed number of rounds in an electro-hydraulically powered turret; direct- and indirect-fire sights are fitted

**Armour:** Welded steel

**Powerplant:** One 912hp (680kW) Detroit Diesel 12V-71QTA diesel engine with 286 Imp gal (1,300 litres) of fuel

**Performance:** Speed, road 43.5mph (70km/h); range, road 342 miles (550km); fording 2ft 11.5in (0.90m) without preparation; gradient 60%; side slope 30%; vertical obstacle 2ft 11.5in (0.90m); trench 9ft 2.25in (2.80m); ground clearance 1ft 6in (0.46m)

**Variant**

**SB 155/39 ATP:** This conventional but comparatively simple self-propelled equipment was under final development in the early 1990s on the basis of a new chassis mounting with the turret accommodating a standard Santa Barbara L/39 howitzer. The turret can be traversed through 360°, and the ordnance can be elevated through an unrevealed arc. The turret has twin loading doors at its rear, and the complete equipment is stabilised for firing by twin jacks, the ammunition range including the full range of NATO types in this calibre. The standard HE projectile can be fired to a maximum range of 26,245yds (24,000m) while the HE rocket-assisted type attains a maximum range of 32,810yds (30,000m). It is possible that final development and production will not occur, as a consequence of the general Western European 'peace dividend' feeling that such expense is unwarranted since the threat from a Soviet-dominated Eastern Europe disappeared.

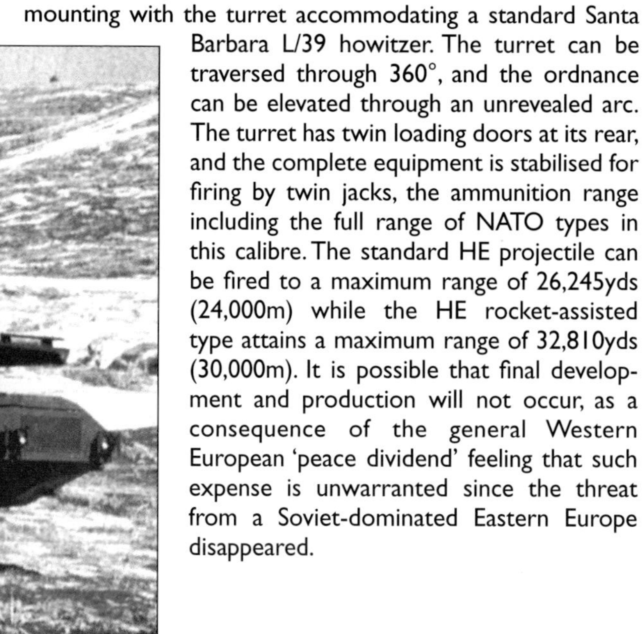

*The SB 155/39 ATP was designed as a self-propelled equipment based on the SB 155/39 towed howitzer that has also been developed in a version with an auxiliary power unit for limited battlefield mobility in the 'shoot and scoot' role.*

# Bofors Bandkanon 1A 155mm SP gun

**Country of origin:** Sweden

**Type:** Tracked self-propelled gun

**Crew:** Five

**Combat weight:** 116,843lb (53,000kg)

**Dimensions:** Length, overall 36ft 1in (11.00m) and hull 21ft 5.875in (6.55m); width 11ft 9.67in (3.37m); height over machine-gun 12ft 7.5in (3.85m)

**Armament system:** One 155mm (6.1in) Bofors gun with 14 rounds of ready-use ammunition and one 7.62mm (0.3in) AA machine-gun in an electrically powered turret; direct- and indirect-fire sights are fitted

**Armour:** Welded steel varying in thickness between 0.39 and 0.79in (10 and 20mm)

*A pioneering self-propelled equipment of its type, the Bandkanon 1A is readily identifiable by its massive external loading system that provides a very high rate of fire and can be reloaded in a mere two minutes.*

**Powerplant:** One 240hp (179kW) Rolls-Royce K60 diesel engine and one 300shp (224kW) Boeing Model 502/10MA gas turbine with 318 Imp gal (1,445 litres) of internal fuel

**Performance:** Speed, road 17.4mph (28km/h); range, road 143 miles (230km); fording 3ft 3.5in (1.00m) without preparation; gradient 60%; side slope 30%; vertical obstacle 3ft 1.5in (0.95m); trench 6ft 6.75in (2.00m); ground clearance 1ft 2.5in (0.37m)

**Variant**

**Bandkanon 1A:** Entering service in 1966 and only with the Swedish army, this self-propelled gun is a massive and slow equipment that uses many automotive components from the Strv 103 MBT, but has an automatic loader that permits a high rate of fire, especially as the loader's associated 14-round clip can be recharged in a mere two minutes. The L/50 ordnance is located in a large turret at the rear of the vehicle, and can be traversed through a total of only 30° (15° left and right of the centreline) above an elevation angle of 0°, the figures declining to a mere 19° (15° left and 4° right of the centreline) below an elevation angle of 0°. The ordnance's elevation limits, within a total arc of 43°, are -3° and +40° in powered mode and -2° to +38° in manual mode.

The 105.8lb (48kg) HE-FRAG projectile is fired at a muzzle velocity of 2,838ft (865m) per second to attain a maximum range of 27,995yds (25,600m).

# T-34/122 122mm SP howitzer

**Country of origin:** Syria

**Type:** Tracked self-propelled howitzer

**Crew:** Six

**Combat weight:** 63,933lb (29,000kg)

**Dimensions:** Length 19ft 8.5in (6.01m); width 9ft 9.75in (2.99m); height 8ft 10.33in (2.70m)

**Armament system:** One 122mm (4.8in) D-30 howitzer with 40 rounds of ready-use ammunition mounted on a manually operated turntable; direct- and indirect-fire sights are fitted

**Armour:** Welded and cast steel

**Powerplant:** One 496hp (370kW) Model V-2 diesel engine with 123 Imp gal (560 litres) of internal fuel

**Performance:** Speed, road 37.3mph (60km/h); range, road 186 miles (300km); fording 4ft 3in (1.30m) without preparation; gradient 35%; side slope not revealed; vertical obstacle 2ft 3.5in (0.70m); trench 7ft 6in (2.29m); ground clearance 1ft 3.75in (0.40m)

**Variant**

**T-34/122:** Introduced in the early 1970s and obsolescent even as it was placed in service, the T-34/122 is a conversion of the elderly T-34 medium tank to mount an exposed 122mm Soviet howitzer in place of the turret, 10 four-round ammunition boxes being attached to the hull sides. The mounting has a practical traverse limit of 120° (60° left and right of the centreline to the rear), and the L/40 ordnance can be elevated through an arc of 77° (-7° to +70°). The projectile types that can be fired are HE-FRAG, HEAT-FS and smoke, the normal complements of these types being 32, four and four respectively. The 47.97lb (21.76kg) HE-FRAG projectile is fired at a muzzle velocity of 2,264ft (690m) per second to attain a maximum range of 16,840yds (15,400m), while the 55.1lb (25kg) HEAT-FS projectile is fired at a muzzle velocity of 2,428ft (740m) per second to penetrate a maximum of 18.1in (460mm) of armour at any range.

The equipment lacks any real credibility on the modern battlefield as it provides no protection for its crew members, who are thus vulnerable to enemy shell fire as well as air attack.

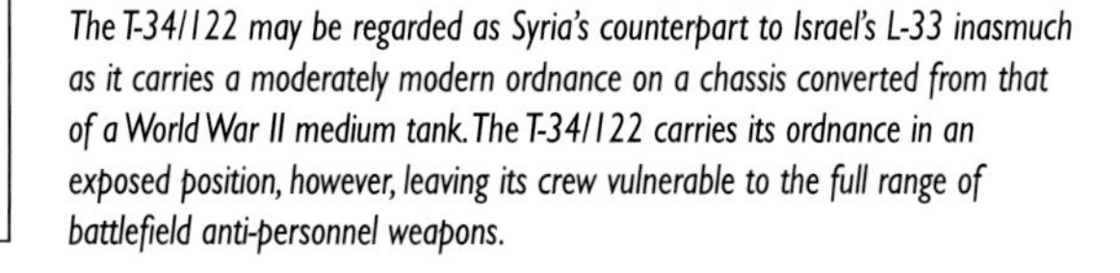

*The T-34/122 may be regarded as Syria's counterpart to Israel's L-33 inasmuch as it carries a moderately modern ordnance on a chassis converted from that of a World War II medium tank. The T-34/122 carries its ordnance in an exposed position, however, leaving its crew vulnerable to the full range of battlefield anti-personnel weapons.*

# 2S4 self-propelled 240mm mortar

**Country of origin:** USSR (now CIS)

**Type:** Tracked self-propelled mortar

**Crew:** Not revealed

**Combat weight:** Not revealed

**Dimensions:** Length, overall not revealed, hull 21ft 2in (6.454m); width 9ft 4.25in (2.85m); height not revealed

**Armament system:** One 240mm (9.45in) M-240 mortar with an unrevealed number of ready-use rounds in a hydraulically powered rear mounting; indirect-fire sights are fitted

**Armour:** Welded steel varying in thickness between 0.28 and 0.55in (7 and 14mm)

**Powerplant:** One 241hp (180kW) YaMZ 238V diesel engine with 99 Imp gal (450 litres) of internal fuel

**Performance:** Speed, road 38.2mph (61.5km/h); range, road 311 miles (500km); fording without preparation not revealed; gradient 60%; side slope 30%; vertical obstacle 2ft 3.5in (0.70m); trench 8ft 10.25in (2.70m); ground clearance 1ft 3.75in (0.40m)

**Variant**

**2S4:** This powerful equipment is based on the chassis of the GMZ tracked minelayer, and carries at its rear the M-240 breech-loading mortar, which is essentially similar to its towed counterpart except for the vehicle mounting attach-ments on the baseplate, allowing the weapon to be pivoted by a hydraulic system round the vehicle rear so that the baseplate is firmly embedded in the ground under the vehicle rear for firing. The mortar can fire HE, chemical, concrete-piercing or tactical nuclear projectiles between minimum and maximum ranges of 875 and 13,890yds

(800 and 12,700m). The HE bomb weighs 286lb (130kg), and the maximum rate of fire is probably one round per minute, possibly with the use of an assisted-loading device. This equipment has the Soviet military and industrial designations SM-240 and 2S4, and in the US system of terminology is known as the M1975.

*The 2S4 is a rare type of weapon inasmuch as it is a self-propelled mortar. Here the 9.45in (240mm) breech-loaded mortar seen as its baseplate is being lowered to the ground by a hydraulic system so that the weapon can be brought into action on a stable platform.*

# 2S7 203mm SP gun

**Country of origin:** USSR (now CIS)

**Type:** Tracked self-propelled gun

**Crew:** Four

**Combat weight:** 88,183lb (40,000kg)

**Dimensions:** Length, overall 42ft 0in (12.80m) and hull 34ft 5.33in (10.50m); width 11ft 5.75 in (3.50m); height 11ft 5.75in (3.50m)

**Armament system:** One 203mm (8in) gun with an unrevealed number of ready-use rounds in a powered limited-traverse mounting; direct- and indirect-fire sights are fitted

**Armour:** Welded steel

**Powerplant:** One 449hp (335kW) diesel engine with an unrevealed quantity of internal fuel

**Variant**
**2S7:** This impressive item of equipment has the Soviet service designation SO-203 and the US designation M1975, and is probably the largest armoured fighting vehicle developed for service with the armies of the Warsaw Pact in the last stages of its existence during the later 1980s. The chassis appears to combine features of the SA-12 self-propelled surface-to-air missile launcher and the T-72 MBT, and has the massive but unprotected gun mounted at its rear with a powered loading system. It is unlikely that more than two ready-use rounds are carried on the vehicle, so the bulk of the equipment's ready-use ammunition supply is carried on an accompanying vehicle. The 2S7 has a large rear-mounted stabilising spade, and its maximum rate of fire is probably two rounds per minute, the sustained rate being one round every two minutes. The weapon is deployed at front (army group) level, and the gun can be elevated to +60° for the delivery of projectiles (220.5lb/100kg conventional and 2/5-kiloton nuclear) to a maximum range of 32,810yds (30,000m).

# 2S5 Giatsint self-propelled 152mm gun

**Country of origin:** USSR (now CIS)

**Type:** Tracked self-propelled gun

**Crew:** Not revealed

**Combat weight:** Not revealed

**Dimensions:** Details not revealed

**Armament system:** One 152.4mm (6in) gun with an unrevealed number of ready-use rounds and one 12.7mm (0.5in) NSV heavy machine-gun in a powered limited-traverse mounting; direct- and indirect-fire sights are fitted

**Armour:** Welded steel varying in thickness between 0.28 and 0.55in (7 and 14mm)

**Powerplant:** One diesel engine of unrevealed power with an unrevealed quantity of internal fuel

**Performance:** Details not revealed

**Variant**
**2S5:** Known in the West by the US designation M1979, this Giatsint (hyacinth) is a powerful equipment based on the chassis of the GMZ minelaying vehicle. The ordnance mounted at the rear of the hull is that of the 2A26 (M1976) towed howitzer, and is able to fire substantial rounds (2/5-kiloton tactical nuclear, chemical, concrete-piercing and HE) to a maximum range of 29,530yds (27,000m). The type can also fire an HE rocket-assisted projectile to a maximum range of about 40,465yds (37,000m).

# 2S3 Akatsiya 152mm SP gun/howitzer

**Country of origin:** USSR (now CIS)

**Type:** Tracked self-propelled gun/howitzer

**Crew:** Six

**Combat weight:** 50,705lb (23,000kg)

**Dimensions:** Length, gun forward 25ft 6.33in (7.78m) and hull 23ft 5in (7.14m); width 10ft 6in (3.20m); height 8ft 11in (2.72m)

**Armament system:** One 152.4mm (6in) modified D-20 gun/howitzer with 46 rounds of ready-use ammunition and one 7.62mm (0.3in) AA machine-gun in a powered turret; direct- and indirect-fire sights are fitted

**Armour:** Welded steel

**Powerplant:** One 523hp (390kW) diesel engine with 110 Imp gal (500 litres) of internal fuel

**Performance:** Speed, road 34mph (54.7km/h); range, road 186 miles (300km); fording 3ft 7.33in (1.10m) without preparation; gradient 60%; side slope 30%; trench 9ft 2.25in (2.80m); ground clearance 1ft 5.75in (0.45m)

**Variant**

**2S3:** Originally known in the West as the M1973, and possessing the Soviet service designation SO-152 and name Akatsiya (acacia), this useful self-propelled gun/howitzer adds the well-tried D-20 ordnance in a neat turret to a chassis/automotive system based on that of the self-propelled launcher for the SA-4 'Ganef' surface-to-air missile system. The type has night vision devices and an NBC protection system, but is not amphibious.

The turret traverses through 360°, and the L/34 ordnance can be elevated through an arc of 68° (-3° to +65°). The ammunition types are HE-FRAG with a 95.9lb (43.5kg) projectile fired to a maximum range of 20,230yds (18,500m), HE RAP with its rocket-assisted projectile fired to a maximum range of 26,245yds (24,000m), HEAT, AP-T (armour-piercing - tracer), smoke, illuminating, and nuclear with a sub-kiloton/5-kiloton selectable-yield warhead. It is believed that an automatic loader may be fitted.

*The 2S3 is a trim equipment based on a well-proved chassis and is capable of firing a large range of projectiles to a tactically useful range. The equipment is provided with an NBC system and night vision devices, but its main tactical limitation is its lack of amphibious capability.*

# 2S1 Gvozdika 122mm SP howitzer

**Country of origin:** USSR (now CIS)

**Type:** Tracked self-propelled howitzer

**Crew:** Four

**Combat weight:** 35,273lb (16,000kg)

**Dimensions:** Length 23ft 11.5in (7.30m); width 9ft 4.25in (2.85m); height 7ft 10.5in (2.40m)

**Armament system:** One 122mm (4.8in) modified D-30 howitzer with 40 rounds of ready-use ammunition in an electrically powered turret; direct- and indirect-fire sights are fitted

**Armour:** Welded steel up to a maximum thickness of 0.79in (20mm)

**Powerplant:** One 241hp (180kW) YaMZ 238 V diesel engine with 121 Imp gal (550 litres) of internal fuel

**Performance:** Speed, road 37.3mph (60km/h) and water 2.8mph (4.5km/h) driven by its tracks; range, road 311 miles (500km); fording amphibious; gradient 60%; side slope 30%; vertical obstacle 3ft 7.25in (1.10m); trench 8ft 10.33in (2.70m); ground clearance 1ft 6in (0.46m)

**Variants**

**2S1:** Originally known in the West as the M1974 and in Soviet service as the SO-122 with the name Gvozdika (carnation), this vehicle uses a modified version of the D-30 ordnance in a new turret on a hull based on that of the MT-LB multi-role vehicle, at least in its chassis and automotive

system. The type is amphibious without any preparation, and has night vision devices and an NBC protection system.

The turret traverses through 360°, and the L/40 ordnance can be elevated through an arc of 73° (-3° to +70°). The main ammunition types are HE-FRAG with a 47.9lb (21.72kg) projectile fired at a muzzle velocity of 2,264ft (690m) per second to a maximum range of 16,730yds (15,300m), HE RAP with a rocket-assisted projectile fired to a maximum range of 23,950yds (21,900m), and HEAT-FS with a 47.7lb (21.63kg) projectile

fired at a muzzle velocity of 2,231ft (680m) per second to penetrate 18.1in (460mm) of armour at any range; other projectile types are smoke and illuminating.

**M1974-1 ACRV:** This is the artillery command and reconnaissance vehicle (ACRV) development of the 2S1 with a large superstructure, and is issued at battalion level. The vehicle is armed with a 12.7mm (0.5in) DShK heavy machine-gun, the crew is five, and the combat weight 30,864lb (14,000kg).

**M1974-2 ACRV:** This is the artillery battery command model with a laser rangefinder.

**M1974-3 ACRV:** This is the battalion staff vehicle with additional communications gear.

**M1974 ACRV/'Big Fred':** This is the version fitted with 'Big Fred' artillery- and mortar-locating radar.

**M1979 MCV:** This is the mineclearing vehicle with a turret and three rockets for the discharge of explosive-filled hose.

*The chassis of the 2S1 is based on major components of the MT-LB tracked multi-role vehicle, and is fully amphibious.*

# 2S9 self-propelled 120mm gun/mortar

**Country of origin:** USSR (now CIS)

**Type:** Tracked airborne-assault self-propelled gun/mortar

**Crew:** Four

**Combat weight:** 19,841lb (9,000kg)

**Dimensions:** Length, gun forward 19ft 9in (6.02m); width 8ft 7.5in (2.63m); height 7ft 6.5in (2.30m)

**Armament system:** One 120mm (4.72in) L/15 gun/mortar with 60 rounds of ready-use ammunition in a powered turret; direct- and indirect-fire sights are fitted

**Armour:** Welded steel varying in thickness between 0.28 and 0.63in (7 and 16mm)

**Powerplant:** One 302hp (225kW) Model 5D-20 diesel engine with an unrevealed quantity of fuel

**Performance:** Speed, road 37.3mph (60km/h) and water 5.6mph (9km/h) driven by two waterjets; range, road 311 miles (500km) and water 56 miles (90km); fording amphibious; gradient 32%; side slope 18%; vertical obstacle 2ft 7.5in (0.80m); trench not revealed; ground clearance variable between 3.94 and 17.7in (0.1 and 0.45m)

**Variant**

**2S9:** Also known in Soviet service terminology as the SO-120, this is an important light support vehicle introduced in 1984 and based on the lengthened chassis of the BMD-2 tracked carrier. The vehicle is designed to provide Soviet airborne formations with powerful direct and indirect fire support, with a turret-mounted 120mm (4.72in) gun/mortar of hybrid type using a clip feed mechanism to generate a maximum rate of fire approaching 30 rounds per minute, the sustained rate being between six and eight rounds per minute. There appears to be no physical limitation to 360° traverse for the turret, the apparent operational limitation to a total of 70° (35° left and right of the centreline) probably reflecting stability problems with the light chassis. The weapon can be elevated through an arc of 84° (-4° to +80°), and is a useful multi-role type able to fire a HEAT projectile capable of penetrating some 23.6in (600mm) of conventional armour at any range. The weapon can also fire indirectly to a range of perhaps 9,625yds (8,800m) with pro-jectiles such as HE, phos-phorus and smoke.

*Light, compact and fitted with a hybrid gun/mortar capable of a very high rate of fire, the 2S9 was designed in the USSR (now CIS) as a support vehicle for airborne forces. The equipment can be delivered to front-line airstrips by tactical transport aircraft, and can also be carried around the battlefield by transport helicopters.*

# Vickers AS-90 155mm SP howitzer

**Country of origin:** UK

**Type:** Tracked self-propelled howitzer

**Crew:** Four

**Combat weight:** 79,365lb (36,000kg)

**Dimensions:** Length, gun forward 31ft 10in (9.70m) and hull 25ft 3in (7.70m); width 10ft 10in (3.30m); height overall 9ft 10in (3.00m)

**Armament system:** One 155mm (6.1in) Royal Ordnance Nottingham howitzer with 40 rounds of ready-use ammunition, and one 7.62mm (0.3in) L37A1 AA machine-gun with 400 rounds in an electrically powered turret; direct- and indirect-fire sights are fitted

**Armour:** Welded steel to a maximum thickness of 0.67in (17mm)

*The AS-90 combines a state-of-the-art ordnance in a fully-powered turret with a capable chassis to create a highly potent battlefield equipment able to deliver high rates of burst fire as the ordnance is provided with a flick rammer.*

**Powerplant:** One 660hp (492kW) Cummins VTA-903T diesel engine with 155 Imp gal (705 litres) of internal fuel

**Performance:** Speed, road 34mph (54.7km/h); range, road 220 miles (354km); fording 4ft 11in (1.50m) without preparation; gradient 60%; side slope 30%; vertical obstacle 2ft 6.5in (0.75m); trench 9ft 10in (3.00m); ground clearance 1ft 4in (0.41m)

**Variant**

**AS-90:** The AS-90 (Artillery System for the 1990s) was selected in 1989 as the British army's replacement for the elderly and undergunned Abbot. The equipment was developed as a private venture by Vickers and the Brazilian company Verolme, the latter being responsible for the chassis, and among the many advanced features of the complete equipment are modest overall weight and dimensions, a high power-to-weight ratio for good performance and battlefield agility, an NBC protection system, night vision devices, and fully automatic fire capability as a result of the advanced fire-control and land-navigation systems.

The turret can be traversed through 360° and the L/39 ordnance can be elevated through an arc of 75° (-5° to +70°). The ordnance is advanced and extremely capable, and is able to fire all types of NATO ammunition in this calibre, while the use of a flick rammer allows a burst rate of three rounds in 10 seconds, or six rounds per minute for a short period, or two rounds per minute indefinitely. The ready-use ammunition supply of 40 rounds is accommodated as 11 rounds in the hull and 29 rounds in the turret bustle, the latter being moved forward into line with the breech by two conveyors. Power-assisted or automatic loading are available as options.

With standard ammunition the ordnance can fire the HE-FRAG projectile to a maximum range of 27,000yds (24,690m) and the ERFB (extended-range full-bore) projectile to a maximum range of more than 35,000yds (32,005m).

# Vickers FV433 Abbot 105mm SP gun

**Country of origin:** UK

**Type:** Tracked self-propelled gun

**Crew:** Four

**Combat weight:** 36,500lb (16,556kg)

**Dimensions:** Length, gun forward 19ft 2in (5.84m) and hull 18ft 1in (5.709m); width 8ft 8in (2.64m); height without AA machine-gun 8ft 2in (2.18m)

**Armament system:** One 105mm (4.13in) L13A1 gun with 40 rounds of ready-use ammunition, one 7.62mm (0.3in) L4A4 AA machine-gun with 1,200 rounds, and three smoke-dischargers on each side of the powered turret; direct- and indirect-fire sights are fitted

**Armour:** Welded steel varying in thickness between 0.24 and 0.47in (6 and 12mm)

**Powerplant:** One 240hp (179kW) Rolls-Royce K60 Mk 4G multi-fuel engine with 85 Imp gal (386 litres) of internal fuel

**Performance:** Speed, road 29.5mph (47.5km/h) and water 3mph (4.8km/h) driven by its tracks; range, road 240 miles (386km); fording 4ft 0in (1.22m) without preparation and amphibious with preparation; gradient 60%; side slope 30%; vertical obstacle 2ft 0in (0.61m); trench 6ft 9in (2.06m); ground clearance 1ft 4in (0.41m)

**Variants**

**FV433 Abbot:** This is a member of the FV430 series of armoured fighting vehicles and uses many of the same automotive components as the FV432 armoured personnel carrier with consequent advantages in training, maintenance and spares holdings. The equipment entered service in 1964, and features night vision devices and NBC protection.

The turret provides 360° traverse, and the comparatively small-calibre ordnance can be elevated through an arc of 75° (-5° to +70°) to fire ammunition types such as HE with a 35.5lb (16.1kg) projectile fired to a maximum range of 18,600yds (17,010m), HESH (HE Squash-Head), smoke and illuminating. The maximum rate of fire is 12 rounds per minute (a rate aided by the provision of a powered rammer), and ammunition is brought up by the Alvis Stalwart 6 x 6 high-mobility truck.

**Value Engineered Abbot:** This is the export version bought by India, and differs in having no night vision or NBC gear, a generally reduced standard of equipment, no AA machine-gun or smoke-dischargers, and a diesel-only 213hp (159kW) version of the K60 engine. This version weighs 35,050lb (15,900kg) and has slightly shorter length.

***Right:*** *One of the smallest-calibre self-propelled equipments developed in the period after World War II, the Abbot is very reliable and can deliver highly accurate fire, but lacks the weight of fire and range of fire required for operations on the modern battlefield.*

***Above right:*** *The Abbot is based on the chassis of the FV432 armoured personnel carrier, offering considerable logistical advantages wherever the self-propelled howitzer and APC are operated together.*

# Pacific Car and Foundry M110A2 8in SP howitzer

**Country of origin:** USA

**Type:** Tracked self-propelled howitzer

**Crew:** Five

**Combat weight:** 62,500lb (28,350kg)

**Dimensions:** Length, gun forward 35ft 2.5in (10.73m) and hull 18ft 9in (5.72m); width 10ft 4in (3.149m); height to top of barrel (travelling position) 10ft 3.75in (3.14m)

**Armament system:** One 8in (203mm) M201 howitzer with two rounds of ready-use ammunition in a hydraulically powered limited-traverse mounting; indirect-fire sights are fitted

**Armour:** Welded steel

**Powerplant:** One 405hp (302kW) Detroit Diesel 8V-71T diesel engine with 216.5 Imp gal (984 litres) of internal fuel

**Performance:** Speed, road 34mph (54.7km/h); range, road 325 miles (523km); fording 3ft 6in

(1.07m) without preparation; gradient 60%; side slope 30%; vertical obstacle 3ft 6in (1.07m); trench 6ft 3in (1.91m); ground clearance 1ft 3.5in (0.39m)

**Variants**

**M110:** This initial model entered American service in 1962 after design and development as a long-range delivery system for massive shells. No NBC protection system was fitted, the ordnance was mounted in an exposed position

above the rear of the vehicle, and only two ready-use rounds were carried: most of the 13-man crew was accommodated in the accompanying M548 support vehicle, which also carried most of the ready-use ammunition. The M2A2 ordnance had a short L/25 barrel producing an overall length of 24ft 6in (7.47m) and a combat weight of 58,500lb (26,535kg), and the mounting over the rear of the hull provided total traverse of 60° (30° left and right of the centreline) together with an elevation arc of 67° (-2° to +65°). A hydraulically powered loading and ramming system was fitted, and ammunition types included HE with a 204lb (92.53kg) projectile fired at a muzzle velocity of 1,925ft (587m) per second to a maximum range of 18,400yds (16,825m), HE grenade-launching, chemical and nuclear (the M422 projectile with 0.5- or 10-kiloton W33 warhead). The standard rate of fire was one round in two minutes.

**M110A1:** This improved model was introduced in 1977 with a longer-barrelled L/37 M201 ordnance for greater propellant load and improved projectile range, including the standard M106 HE projectile fired at a muzzle velocity of 2,333ft (711m) per second to a maximum range of 23,300yds (21,305m). In this model the fuel capacity is reduced from the 250 Imp gal (1,136 litres) of the M110,

thus reducing range, although the extra weight of the longer ordnance also adversely affected speed, trench-crossing capability and ground clearance (in which aspects the original M110 had equalled the M107).

**M110A2:** This was introduced in 1978 as the definitive version of the M110A1 with a double-baffle muzzle brake. The equipment fires the same rounds as the M110 together with an HE rocket-assisted projectile fired to a maximum range of 31,825yds (29,100m) and the M753 nuclear RAP with a 0.5-, 1- or 2-kiloton selectable yield W79-1 warhead fired to the same maximum range.

**M578:** This is the armoured recovery vehicle of the series that uses the same chassis and automotive system but is fitted with a winch and other specialised equipment.

The M110 and M578 series has been used by Belgium, Germany, Greece, Iran, Israel, Italy, Japan, Jordan, Netherlands, Pakistan, Saudi Arabia, South Korea, Spain, Taiwan, Turkey, the UK and the USA.

*An obsolescent type because of its exposed ordnance and lack of crew protection, the M110 fires a potent range of projectiles to long range.*

# **P**acific **C**ar and **F**oundry **M**107 175mm **S**P gun

**Country of origin:** USA

**Type:** Tracked self-propelled gun

**Crew:** Five

**Combat weight:** 62,100lb (28,168kg)

**Dimensions:** Length, gun forward 36ft 11in (11.26m) and

*The M107 is another obsolescent equipment offering no protection for its ordnance and crew, but is still valued by some operators for its ability to deliver potent projectiles over considerable ranges with excellent accuracy.*

hull 18ft 9in (5.72m); width 10ft 4in (3.15m); height to top of barrel (travelling position) 12ft 1in (3.68m)

**Armament system:** One 175mm (6.89in) M113 gun with two rounds of ready-use ammunition in a hydraulically powered limited-traverse mounting; direct- and indirect-fire sights are fitted

**Armour:** Welded steel

**Powerplant:** One 405hp (302kW) Detroit Diesel 8V-71T diesel engine with 250 Imp gal (1,136 litres) of internal fuel

**Performance:** Speed, road 35mph (56.3km/h); range, road

450 miles (724km); fording 3ft 6in (1.07m) without preparation; gradient 60%; side slope 30%; vertical obstacle 3ft 6in (1.07m); trench 7ft 6in (2.30m); ground clearance 1ft 6.25in (0.47m)

### Variant

**M107:** This very substantial and unwieldy equipment was designed as the partner for the M110 howitzer, and was intended specifically for the delivery of heavy fire over very long ranges. One of the main features of the system was the use of identical chassis and ordnance mounting, allowing the rapid change of the M107 into the M110 and vice versa. The type entered service in 1962.

The ordnance mounting over the rear of the hull

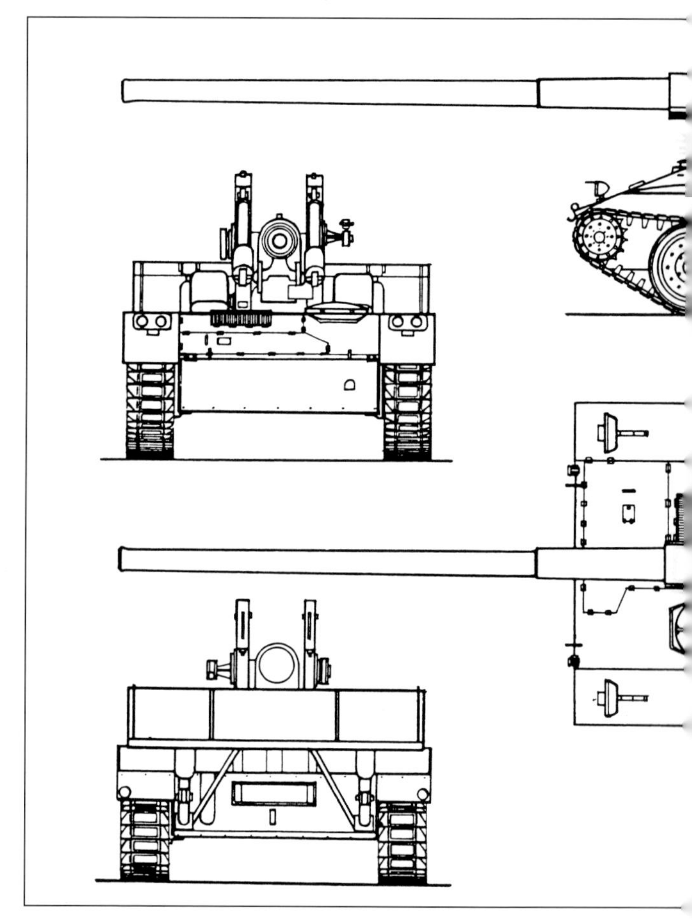

provides for a total traverse of 60° (30° left and right of the centreline), and the L/60 ordnance can be elevated through an arc of 67° (-2° to +65°), firing a 147.25lb (66.8kg) HE projectile at a muzzle velocity of 2,992ft (912m) per second to a maximum range of 35,750yds (32,690m). Like the M110, the M107 is stabilised for firing by two rear-mounted spades, and normally fires at the rate of one round every two minutes, although the rate of one round per minute can be achieved for short periods, due to the provision of a hydraulically powered loading and ramming system. The gun crew and the bulk of the ready-use ammunition are carried in an M548 support vehicle.

The M107 has been operated by Germany, Greece, Iran, Israel, South Korea, Spain, Turkey, the UK and the USA.

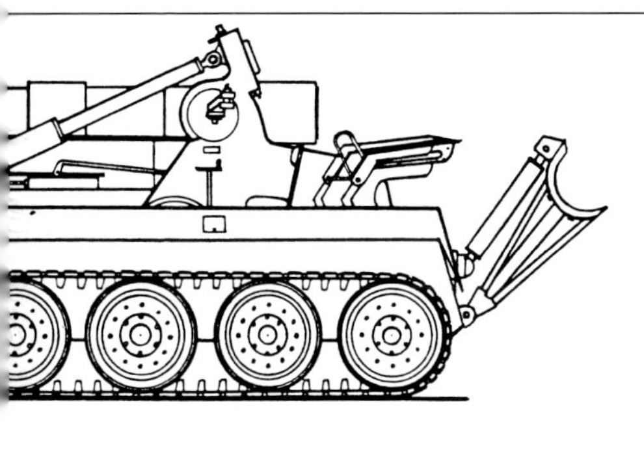

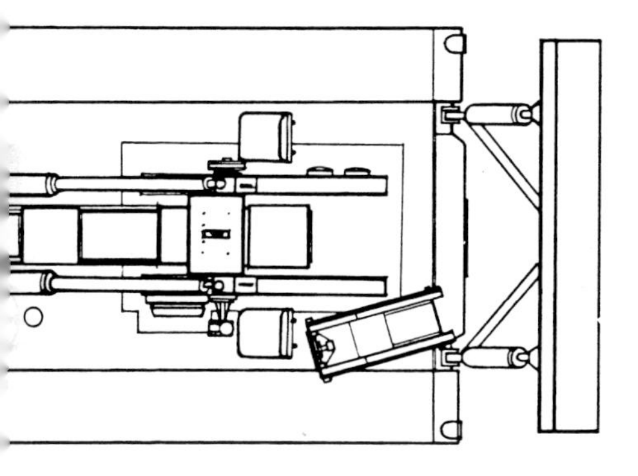

*Key features of the M107's layout are the guns' very long barrel for excellent range and accuracy factors, and the rear-mounted hydraulically operated twin spades, which are lowered into the ground to stabilise the equipment in firing position.*

# Cleveland Army Tank Plant/Bowen-McLaughlin-York M109A2 155mm SP howitzer

**Country of origin:** USA

**Type:** Tracked self-propelled howitzer

**Crew:** Six

**Combat weight:** 55,000lb (24,948kg)

**Dimensions:** Length, gun forward 29ft 11in (9.12m) and hull 20ft 3.75in (6.19m); width 10ft 2in (3.10m); height including machine-gun 10ft 9in (3.28m)

**Armament system:** One 155mm (6.1in) M126 howitzer with 36 rounds of ready-use ammunition, and one 0.5in (12.7mm) Browning M2HB AA heavy

*The M109 series is the Western world's single most important item of self-propelled artillery, offering great flexibility of action with a high degree of reliability and ready upgrade potential.*

machine-gun with 500 rounds in a hydraulically powered turret; direct- and indirect-fire sights are fitted

**Armour:** Welded aluminium

**Powerplant:** One 405hp (302kW) Detroit Diesel 8V-71T diesel engine with 112 Imp gal (511 litres) of internal fuel

**Performance:** Speed, road 35mph (56.3km/h); range, road

215 miles (346km); fording 3ft 3in (1.00m) without preparation; gradient 60%; side slope 40%; vertical obstacle 1ft 9in (0.53m); trench 6ft 0in (1.83m); ground clearance 1ft 5.75in (0.45m)

**Variants**

**M109:** This type entered service late in 1962 at the beginning of a programme that has seen the series being extensively built and developed as successor to the M44 series, with greater operational capabilities combined with better crew protection. The basic model weighed 44,200lb (20,049kg) and had IR driving lights but no NBC protection system. The fording capability was 6ft 0in (1.83m) without preparation, although the addition of nine air bags gave the type a limited amphibious capability at a track-propelled water speed of 4mph (6.4km/h).

The turret was fitted at the rear of the vehicle and provided 360° traverse, while the L/20 ordnance could be elevated through an arc of 78° (-3° to +75°). The vehicle was stabilised for firing by two rear-mounted spades, and the onboard ammunition stowage amounted to 28 rounds. The ammunition types available included HE with a 94.6lb (42.91kg) projectile fired at a muzzle velocity of 1,845ft (562m) per second to a maximum range of 16,000yds (14,630m), HE grenade- and mine-launching, HE rocket-assisted, illuminating, smoke, chemical, M454 nuclear (with a 0.1/1-kiloton W48 warhead) and laser-guided Copperhead (two of these 22,000yd/20,110m-range cannon-launched guided projectiles generally being carried in place of other rounds). The standard rate of fire was one round per minute, although three rounds per minute could be achieved.

**M109A1:** This was introduced as an improved M109 with longer L/33 M185 ordnance and a new charge system extending HE range to 19,800yds (18,105m) at a muzzle velocity of 2,245ft (684m) per second, and RAP range to 26,250yds (24,005m). The type was produced by M109 conversion, and entered service in 1973 as a variant with a weight of 53,065lb (24,070kg) and an overall length of 29ft 8in (9.04m).

**M109A2:** Entering service in 1979, this is the definitive M109 model based on the M109A1 but with many detail improvements over earlier types and with a turret bustle for greater ammunition capacity. The version proposed for

Thailand by BMY and Rheinmetall is the M109A2T, with a new ordnance combining features of the FH-170 and M198 towed howitzers to provide a range of 27,000yds (24,690m) with the standard projectile and 32,810yds (30,000m) with the base-bleed projectile.

**M109A3:** This designation is used for earlier vehicles brought up to M109A2 standard.

**M109A4:** This designation is used for earlier vehicles upgraded in the Howitzer Extended Life Program (HELP) with the RAM-D (reliability, availability, maintainability - durability) package, improved battlefield survivability features and the addition of an NBC protection system.

**M109A5:** This designation is used for M109A1 and M109A2 vehicles upgraded from the late 1980s in the Howitzer Improvement Program (HIP), with the RAM-D package of reliability developments, plus extended range, an automatic fire-extinguishing system, an aluminium/Kevlar turret, a full-width turret bustle with storage for 36 charges, an ordnance loading-assist device, and an improved fire-control system (including an onboard ballistic computer and inertial reference system). The nature of the ordnance is as yet uncertain, the choice still lying between the current M185 or a new ordnance with an L/39 or L/53 barrel for significantly improved range. The type is capable of firing a new nuclear round with a W82 warhead.

**M109A6:** Under development as a joint Israeli-American programme based on the M109A5, this variant offers the ability to engage targets at a range of 33,000yds (30,175m) and to fire the first aimed round within 30 seconds of the vehicle coming to a halt. The type has a Honeywell automatic fire-control system and an onboard navigation system.

**M109AL:** This is an Israeli M109A1 version with stowage racks to permit the external carriage of personal kit, so boosting internal ammunition capacity

**M109G:** This is a German version of the M109 with a revised breech boosting range to 20,230yds (18,500m) with the standard HE projectile, a new muzzle brake, tracks with features of the Leopard 1's tracks, a 7.62mm (0.3in) MG3 AA machine-gun, three smoke-dischargers on each side of the turret, and German fire-control equipment. Existing equipments are being upgraded to M109A3G standard with a new L/39 ordnance based on that of the FH-70 towed howitzer, which allows the full range of FH-70 and NATO projectiles to be fired, the standard HE type reaching a maximum range of 27,010yds (24,700m) and the HE base-bleed type attaining 32,810yds (30,000m). The new ordnance can also fire the latest types of carrier projectiles, and is supplied from a new projectile magazine in the turret bustle: this has outward-opening doors for reloading purposes, and accommodates 22 of the 34 projectiles carried.

**M109L:** This is an Italian M109 version carrying an OTO Melara L/39 ordnance with a double-baffle muzzle brake (essentially that of the ordnance used in the M109G) and fume extractor, and able to fire the ammunition of the FH-70 howitzer to ranges of between 26,245yds (24,000m) with standard projectiles and 32,810yds (30,000m) with RAPs. Other modifications include an armoured hood for the roof-mounted sight, and a revised bustle to allow the carriage of extra ammunition.

*The ordnance of the M109, like that of other self-propelled artillery equipments, is held in a travelling lock when the vehicle is on the move.*

**M109U:** This is a Swiss version, the original Panzerhaubitze 66 type being the M109, which was followed by the Panzerhaubitze 74 (M109A1) and Panzerhaubitze 66/74 (Pzh 66 upgraded to Pzh 74 standard). These Swiss vehicles have a revised electrical system and a loading system that operates at all angles of ordnance elevation.

**M109 (Netherlands):** This is a turretless version used by the Dutch army for driver training.

**M109 (Norway):** The Norwegian army's fleet of M109 self-propelled howitzers is being upgraded by Norsk Forsvarateknologi with a new L/37 ordnance (supplied by Rheinmetall) and other modifications to improve battlefield capability.

**M109 (Taiwan):** Under the local designation XT-69, Taiwan has combined the chassis and automotive system of the M109 with an exposed mounting for an ordnance derived from the US 155mm (6.1in) M114 towed howitzer but fitted with a multi-baffle muzzle brake. The crew is five, and it is believed that some 25 rounds of ammunition are

*One of the keys to the tactical versatility of the M109 series' variants is the ability of the ordnance to fire the full range of NATO 155mm (6.1in) artillery rounds as well as laser-guided projectiles and a number of special rounds developed for national use by members of the NATO alliance.*

carried, the ordnance having a maximum range of some 16,405yds (15,000m).

**M992:** This is the field artillery ammunition support vehicle derived from the M109, with the turret replaced by a large superstructure containing an extending conveyor system for the movement of 155mm (6.1in) ammunition. Stowage amounts to 93 projectiles, 99 propellant charges and 104 fuses. The type can also be used for 175 and 203mm (6.89 and 8in) ammunition, stowage of the latter being 98 projectiles, 53 charges and 56 fuses. It has also been proposed that the M992 could be used as the basis of a whole family of self-propelled artillery support vehicles, including fire-direction centre, command-post, ambulance, rearmament, and maintenance assistance vehicles.

Operators of the M109 and M992 series have included Austria, Belgium, Canada, Denmark, Egypt, Ethiopia, Germany, Greece, Iran, Israel, Italy, Jordan, Kuwait, Morocco, Netherlands, Norway, Pakistan, Peru, Portugal, Saudi Arabia, South Korea, Spain, Switzerland, Taiwan, the UK and the USA.

# Massey Harris M44A1 155mm SP howitzer

**Country of origin:** USA

**Type:** Tracked self-propelled howitzer

**Crew:** Five

**Combat weight:** 64,000lb (29,030kg)

**Dimensions:** Length 20ft 2.5in (6.16m); width 10ft 7.5in (3.24m); height over tarpaulin cover 10ft 2.5in (3.11m)

**Armament system:** One 155mm (6.1in) M45 howitzer with 24 rounds of ready-use ammunition, and one 0.5in (12.7mm) Browning M2HB AA heavy machine-gun with 900 rounds in a manually powered limited-traverse mounting; direct- and indirect-fire sights are fitted

**Armour:** Welded steel to a standard thickness of 0.5in (12.7mm)

**Powerplant:** One 500hp (373kW) Continental AOSI-895-5 petrol engine with 125 Imp gal (568 litres) of internal fuel

**Performance:** Speed, road 35mph (56.3km/h); range, road 82 miles (132km); fording 3ft 6in (1.07m) without preparation; gradient 60%; side slope not revealed; vertical obstacle 2ft 6in (0.76m); trench 6ft 0in (1.83m); ground clearance 1ft 7in (0.48m)

**Variants**

**M44A1:** This weapon uses the same basic chassis and automotive system as the 105mm (4.13in) M52, which are very similar to those of the M41 light tank, and the result is the last US self-propelled artillery equipment with an open-topped fighting compartment. The original M44 had a normally aspirated engine, but was replaced in the mid-1950s by the M44A1 with a fuel-injected engine. The turret has a total traverse of 60° (30° left and right of the centreline) and the L/20 ordnance can be elevated through an arc of 70° (-5° to +65°). The ammunition types are HE, HE grenade-launching, smoke, illuminating and chemical, and maximum range with the 94.6lb (42.91kg) HE projectile is 16,000yds (14,630m). The normal rate of fire is one round per minute with the vehicle stabilised by two rear-mounted spades.

**Rheinmetall M44:** This is the German upgrade of the original weapon. The revised vehicle is powered by a 450hp (336kW) MTU 883 Aa diesel engine for a range of 373 miles (600km), and has many detail modifications. The main armament is a Rheinmetall 155mm (6.1in) L/39 howitzer based on the FH-70 ordnance, and the mounting allows ordnance elevation through an arc of 70° (-5° to +65°). Some 30 rounds of ammunition can be carried, and maximum projectile ranges are 26,245yds (24,000m) with standard ammunition and 32,810yds (30,000m) with extended-range ammunition.

*The M44 is an obsolescent equipment based, like the M52 with its smaller-calibre 105mm (4.13in) howitzer, on a development of the M41 light tank's chassis.*

# Cadillac (General Motors) M108 105mm SP howitzer

**Country of origin:** USA

**Type:** Tracked self-propelled howitzer

**Crew:** Five

**Combat weight:** 49,500lb (22,453kg)

**Dimensions:** Length 20ft 0.75in (6.11m); width 10ft 9.75in (3.295m); height including machine-gun 10ft 4.25in (3.155m)

**Armament system:** One 105mm (4.13in) M103 howitzer with 87 rounds of ready-use ammunition, and one 0.5in (12.7mm) Browning M2HB AA heavy machine-gun with 500 rounds in a manually powered turret; direct- and indirect-fire sights are fitted

**Armour:** Welded aluminium

**Powerplant:** One 405hp (302kW) Detroit Diesel 8V-71T diesel engine with 112 Imp gal (511 litres) of internal fuel

**Performance:** Speed, road 35mph (56.3km/h) and water 4mph (6.4km/h) driven by its tracks; range, road 240 miles (386km); fording 6ft

0in (1.83m) without preparation and amphibious with preparation; gradient 60%; side slope not revealed; vertical obstacle 1ft 9in (0.53m); trench 6ft 0in (1.83m); ground clearance 1ft 5.75in (0.45m)

**Variant**

**M108:** Introduced to US service in 1964, the M108 shares a common chassis and automotive system with the M109 155mm (6.1in) self-propelled howitzer, and also can be fitted with flotation bags to provide a measure of amphibious capability. The type lacks an NBC protection system, but has IR driving lights. The turret traverses through 360°, and the ordnance can be elevated through an arc of 78° (-4° to +74°). The ammunition types available are HE, HE grenade-launching, illuminating, smoke and chemical, and typical performance includes a maximum range of 12,600yds (11,520m) with an HE round that weighs 39.9lb (18.1kg) complete.

# Detroit Arsenal M52A1 105mm SP howitzer

**Country of origin:** USA

**Type:** Tracked self-propelled howitzer

**Crew:** Five

**Combat weight:** 53,000lb (24,041kg)

**Dimensions:** Length 19ft 0in (5.80m); width 10ft 4in (3.15m); height including machine-gun 10ft 10.5in (3.32m)

**Armament system:** One 105mm (4.13in) M49 howitzer with 102 rounds of ready-use ammunition, and one 0.5in (12.7mm) Browning M2HB AA heavy machine-gun with 900 rounds in a manually powered turret; direct- and indirect-fire sights are fitted

**Armour:** Welded steel to a standard thickness of 0.5in (12.7mm)

**Powerplant:** One 500hp (373kW) Continental AOSI-895-5 petrol engine with 150 Imp gal (681 litres) of internal fuel

**Performance:** Speed, road 42mph (67.6km/h); range, road 100 miles (161km); fording 4ft 0in (1.22m) without preparation; gradient 60%; side slope not revealed; vertical obstacle 3ft 0in (0.91m); trench 6ft 0in (1.83m); ground clearance 1ft 7.33in (0.49m)

**Variant**

**M52A1:** This is the 105mm (4.13in) equivalent of the 155mm (6.1in) M44, introduced in 1954 and possessing the same basic characteristics apart from its primary armament and associated ammunition in a turret capable of 360° traverse. The ordnance can be elevated through an arc of 75° (-10° to +65°), and the ammunition types include HE with a projectile fired to a maximum range of 12,325yds (11,270m), HEAT, HE grenade-launching, smoke, illuminating and chemical. The original normally aspirated AOS-895-3 engine of the M52 was soon supplanted by the fuel-injected AOSI-895-5.

This equipment has no need of rear-mounted spades to stabilise it in the firing position. The M52A1 has been used by Greece, South Korea, Spain and the USA.

*The M52 is an obsolete equipment offering only limited tactical capability on the modern battlefield.*